11th AIChE Southwest Process Technology Conference 2019

Houston, Texas, USA
1 - 2 October 2019

ISBN: 978-1-7138-0453-6

Printed from e-media with permission by:

Curran Associates, Inc.
57 Morehouse Lane
Red Hook, NY 12571

Some format issues inherent in the e-media version may also appear in this print version.

Printed with permission by Curran Associates, Inc. (2020)

For permission requests, please contact AIChE
at the address below.

AIChE
120 Wall Street, FL 23
New York, NY 10005-4020

Phone: (800) 242-4363
Fax: (203) 775-5177

www.aiche.org

Additional copies of this publication are available from:

Curran Associates, Inc.
57 Morehouse Lane
Red Hook, NY 12571 USA
Phone: 845-758-0400
Fax: 845-758-2633
Email: curran@proceedings.com
Web: www.proceedings.com

TABLE OF CONTENTS

CHEMICAL PROCESS TECHNOLOGY

DISTILLATION & SEPARATION

PROCESS SAFETY

2-OCT-19

CLIMATE SOLUTIONS

The Energy Transition

Conference: Southwest Process Technology Conference

Year: 2019

Proceeding: 11th AIChE Southwest Process Technology Conference

Group: Southwest Process Technology Conference

Session: Tuesday Keynote Address

Time: Tuesday, October 1, 2019 - 8:15am-8:40am

Authors:

Hofmeister, J., Citizens for Affordable Energy

Keynote Speaker: John Hofmeister

The Energy Transition

We are confronted with an unprecedented challenge. Our planet is warming at an unacceptable rate. Over the next few decades, we must transition responsibly from our current energy mix to one that works to achieve net zero global emissions. In this transition, what are the challenges that confront our industry? What will impact our companies in terms of governance, public policy, reputation, valuation, and societal acceptance? Impacts on the economy, careers and society must be taken into account in the strategy and decision making processes at the highest levels of corporate boards and executive committees, and their relationships with governments and other stakeholders, such as NGOs and customers. Dysfunctionality of todayâ€™s US governance system is a sober concern as it affects long range planning of energy supply, infrastructure and the environment. Are there opportunities as well as challenges? Yes.

Leveraging Hydrocracking Flexibility to Improve Refinery Margins

Conference: Southwest Process Technology Conference

Year: 2019

Proceeding: 11th AIChE Southwest Process Technology Conference

Group: Southwest Process Technology Conference

Session: Catalysis & Reaction Engineering

Time: Tuesday, October 1, 2019 - 9:30am-10:00am

Authors:

Chopra, V., Sell Catalyst & Technologies

Abstract:

The global refining market continues to be very challenging, as oil prices remain low and economic drivers push refiners to process heavier, more sour, and more difficult crudes. These opportunity crudes generate a higher percentage of the distressed streams which are either blended into lower value dispositions or are fed to different conversion units in the refinery. Upgrading these difficult streams to more valuable liquid products becomes paramount under the current environment as the low-value dispositions continue to become unavailable in the future â€“ especially considering the recent International Maritime Organization (IMO) decision on bunker fuel quality requirements.

Leveraging the flexibility of hydrocracker pretreat catalyst systems to upgrade low value, highly processed streams via hydrocracking without sacrificing run-length, product quality, or margin by disposition as low-valued exports can improve refinery margins. The following case studies offer an exploration that showcases ways to leverage the flexibility of the hydrocracker to process various difficult streams. These streams include, but are not limited to, products from thermal cracking, de-asphalting units, catalytic cracking, coking, lube extraction, or ebullated bed residue upgraders. The three case studies described here demonstrate the different operating challenges and catalyst strategies employed to upgrade a broad range of very difficult feedstocks into premium quality clean fuel products.

The 3rd case study focuses on Shellâ€™s Scotford refinery and illustrates the evolution of pretreat design and operating strategies that address the increasingly demanding cycle objectives for converting synthetic feeds derived from tar sand bitumen via hydrocracking. It also emphasizes the value that close working of both refiner and catalyst supplier can bring by identifying opportunities to improve unit performance over multiple cycles and creating solutions to improve refinery margins.

Time-Accurate, Three-dimensional Simulation of Agitated Vessels

Conference: Southwest Process Technology Conference

Year: 2019

Proceeding: 11th AIChE Southwest Process Technology Conference

Group: Southwest Process Technology Conference

Session: Catalysis & Reaction Engineering

Time: Tuesday, October 1, 2019 - 10:00am-10:30am

Authors:

Thomas, J. A., M-Star Simulations

Abstract:

In many agitated tanks and fluid handling systems, yield and performance are governed by complex turbulent fluid mechanics. These turbulent motions, which can inform both bulk transport and reaction kinetics, occur over a large range of length and time-scales. From a modeling perspective, direct numerical simulation of this entire turbulence spectrum is required to obtain complete insights into systems performance. Due to time and resource constraints, however, such detailed simulations have not historically been practical within most industrial settings.

In this work, we show how graphical processing units (GPUs) can make direct numerical simulation (DNS) of industrial mixing systems practical and timely. Although GPUs have historically been used for image rendering, in recent years they have emerged and as powerful computational rivals to traditional CPUs. We will begin the presentation by introducing the concepts governing fluid modeling on GPUs. We then show how, given identical algorithms, a single scientific GPU can execute simulations two to three orders-of-magnitude faster than a single CPU. We then present various criteria for monitoring the convergence from large-eddy simulation (LES) to direct numerical simulation, within the context of the turbulent fluid motion. We then apply these criteria to study blending, energy dissipation, and reaction rates in a benchtop reactor.

Applications of Catalyst Enhancement Solutions in Oil Refining Processes

Conference: Southwest Process Technology Conference

Year: 2019

Proceeding: 11th AIChE Southwest Process Technology Conference

Group: Southwest Process Technology Conference

Session: Catalysis & Reaction Engineering

Time: Tuesday, October 1, 2019 - 11:00am-11:30am

Authors:

Oyekan, S. O., Prafis Energy Solutions

Applications of Catalyst Enhancement Solutions in Oil Refining Processes

By

Soni O. Oyekan

President/CEO

Prafis Energy Solutions

Richmond, TX 77407

Catalysts accelerate the rates of chemical reactions, enable catalytic processes to attain excellent product selectivities, productivities, and profitability in petrochemical and oil refining processes. Catalysis and reaction engineering have served us excellently as correct and effective usage of catalysts have led to significant gains in the productivity and profitability of oil refining processes. Since oil refining units process a wide range of crude oil fractions, a variety of heterogeneous catalysts and catalytic additives are used in oil refining processes to meet a wide range of process objectives and goals.

A short introduction of catalyst compositions, relevant physical and chemical properties of some selected heterogeneous catalysts will be given for some naphtha processes. Process technology driven principles and catalytic enhancing strategies and concepts would be reviewed. Specific examples of applications of catalysts, and additives for catalyst activation, catalyst startup, and feed processing will be discussed. Catalyst enhancing strategies and applications would be covered for a platinum mordenite catalyzed C8 aromatics isomerization process, platinum aluminum chloride catalyzed paraffin isomerization processes, and for platinum catalyzed naphtha reforming processes. Strategies for optimal use of the multi-functional platinum catalysis in naphtha reforming processes would also be covered. Some comments would be offered with respect to the use of catalysts and additives in the processing of gas oils in Fluid Catalytic Cracking units (FCCUs).

Honeywell UOP UniflexTM MC Slurry Hydrocracking Technology

Conference: Southwest Process Technology Conference

Year: 2019

Proceeding: 11th AIChE Southwest Process Technology Conference

Group: Southwest Process Technology Conference

Session: Catalysis & Reaction Engineering

Time: Tuesday, October 1, 2019 - 11:30am-12:00pm

Authors:

Sun, P., Honeywell

High sulfur bunker fuel will be restricted from use in ships without scrubbers after 1st Jan 2020. This is expected to significantly reduce demand for high sulfur fuel oil (HSFO). Refineries with significant HSFO production will require profitable residue conversion projects. For customers targeting zero HSFO, the Uniflex MC Process enables HSFO conversion with an excellent payback. The Uniflex MC Process also provides a good opportunity for refiners shifting from fuels to petrochemicals production.

This presentation summarizes the key process features of Honeywell UOPâ€™s Uniflex MC technology. A unique application of molybdenum-based Microcat catalyst provides stable operation at elevated severity. The Uniflex MC process with novel recycle scheme and solids handling unit delivers high resid conversion, as well as excellent yield structure. Further integration with in-line hydroprocessing results in even higher economic returns. A few case studies illustrate how customers can utilize Honeywell UOPâ€™s slurry hydrocracking solution to make the best of their investment.

Energy Optimization Early in the Design Process Minimizes Process Energy Consumption and Minimizes Total Projects Costs

Conference: Southwest Process Technology Conference

Year: 2019

Proceeding: 11th AIChE Southwest Process Technology Conference

Group: Southwest Process Technology Conference

Session: Energy Efficiency

Time: Tuesday, October 1, 2019 - 9:30am-10:00am

Authors:

McMullan, A., KBC Advanced Technologies, Inc.
Rudman, A., KBC Advanced Technologies, Inc.
Severson, D., KBC Advanced Technologies, Inc.

The typical project lifecycle includes only limited energy and utility analysis during early design, and that usually occurs near the end of FEED. This is often because the team lacks good input data before the design is quantified. As designs progress, there is more data, but process changes to reduce energy consumption become more expensive.

Incorporating optimization studies while the design is still flexible helps to yield maximum benefits for energy and capital minimization. The lack of process data does not have to be a roadblock to earlier optimization. Process Simulation, Pinch Analysis, Utility System Modeling and Experience and Expertise can help fill the gaps in the information needed.

By doing this analysis as early as possible in the design phase, maximum energy consumption reduction and capital outlay reduction benefits can be obtained. The cost impact of identified design changes will be much less than changes later in the design process, as most of the improvements are included in the initial design.

The presentation highlights how to incorporate energy efficiency earlier in the design process and provides a case study to show the benefits of this approach.

Air Preheat System Upgrade on Coker Heaters

Conference: Southwest Process Technology Conference

Year: 2019

Proceeding: 11th AIChE Southwest Process Technology Conference

Group: Southwest Process Technology Conference

Session: Energy Efficiency

Time: Tuesday, October 1, 2019 - 10:00am-10:30am

Authors:

Garg, A., Furnace Improvements Services

One of the refineries in the gulf coast region had two coker heaters that were designed to operate on balanced draft. The heaters had separate air preheater (APH) systems with FD and ID fans. The heaters were connected to a common grade mounted stack. The client was limited on the charge rate when operating the air preheaters. The existing APHs were leaking, and as a result both the FD and ID fans were limiting. The client was by-passing approximately 50% of the hot flue gas directly to the stack. The heaters were also facing O_2 limitations. The client was getting more capacity on natural draft but losing a lot of energy to the stack. FIS conducted a study and identified the deficiencies in the air preheater system.

New APHs were designed for higher heat transfer duty than the existing APHs, and new FD and ID fans were redesigned for higher capacities than the existing fans. A new steam preheat coil was installed at the inlet of the main APH. The steam preheat coils and combustion air by-pass ducts were designed to minimize the acid dew point corrosion in the APH cold blocks. The new combustion air by-pass ducts were sized for 100% by-pass of combustion air. This would help run the heater on forced draft when the APHs get corroded. The existing convection section had two future row provisions. Two rows of process coils were installed to maximize the heater efficiency. The heaters have been commissioned successfully and have been operating for the past six months.

Efficient Pollution-Free Steam Generation with Medium Voltage Electric Boilers

Conference: Southwest Process Technology Conference

Year: 2019

Proceeding: 11th AIChE Southwest Process Technology Conference

Group: Southwest Process Technology Conference

Session: Energy Efficiency

Time: Tuesday, October 1, 2019 - 11:00am-11:30am

Authors:

Taylor, D., Chromalox Inc.

Steam has always been the primary heating source for the chemical processing industries. Historically the steam for process heating was generated from coal, oil, and other fossil fuel sources. Due to large heat duties and steam flow rates required it has not been feasible to consider low voltage electric steam generators since many applications require megawatts of power. However, now with medium voltage resistance heating, large multi-MW electric boilers can be installed and operated at much lower costs than low voltage electric boilers. When deciding which type of steam generation source to utilize it is critical to consider the total cost involved with purchasing, installing, operating, and maintaining your process heating equipment.

This presentation highlights the benefits of utilizing electric power to generate steam compared to traditional fossil fuel and low voltage electric steam generation sources. Advancements in medium voltage heating technology now allows electric process heating to be a viable heat source for large heat duty applications. Utilizing DirectConnect medium voltage technology (up to 7200V) reduces amp draw when compared to low voltage (480V). This reduction in amp draw drastically reduces the electrical installation costs. In specific applications, electric and fired heat sources can be utilized in tandem for optimal efficiency.

With the increased pollution reduction requirements and availability of renewable power sources, electric steam generation can reduce on-site costs by increasing efficiency and eliminating pollution sources.

Address Distillation Process Control During Design Phase to Save Energy and Increase Capacity

Conference: Southwest Process Technology Conference

Year: 2019

Proceeding: 11th AIChE Southwest Process Technology Conference

Group: Southwest Process Technology Conference

Session: Energy Efficiency

Time: Tuesday, October 1, 2019 - 11:30am-12:00pm

Authors:

Herzog, C., Herzog Process Services

Through a series of examples, this paper examines strategies to address basic process control issues during the engineering design phase to increase system capacity and improve on-spec performance. It also presents a view of process control as a life-cycle discipline, rather than an activity that begins after startup; and it shows how improved process control often has a powerful impact on energy efficiency â€" even when you donâ€™t expect it.
Through a series of examples, this paper examines strategies to address basic process control issues during the engineering design phase to increase system capacity and improve on-spec performance. It also presents a view of process control as a life-cycle discipline, rather than an activity that begins after startup; and it shows how improved process control often has a powerful impact on energy efficiency â€" even when you donâ€™t expect it.

Intelligent Processes: Leveraging the Data Revolution for Decision and Control

Conference: Southwest Process Technology Conference

Year: 2019

Proceeding: 11th AIChE Southwest Process Technology Conference

Group: Southwest Process Technology Conference

Session: IIoT/Big Data

Time: Tuesday, October 1, 2019 - 9:30am-10:00am

Authors:

Gopalunni, B., University of British Columbia

We are currently at the dawn of what is considered the fourth industrial revolution. This revolution is driven by a combination of internet connected devices, large volumes of data, significant computing power and major algorithmic advances in artificial intelligence. It presents a unique opportunity for process industries to interpret data in new ways and derive information necessary to enable intelligent decisions at various levels of operation.

We will start with an introduction to major recent breakthroughs in artificial intelligence and describe how they can be used to solve problems in process modeling, fault detection and diagnosis and control. In particular, we will take the audience through a journey of process data analytics that includes traditional multivariate statistical approaches as well as more modern tools such as variational auto encoders and deep neural networks. This will be followed by a quick tour of reinforcement learning and its potential to change the decision and control paradigm as we know it today. The talk will be interspersed with several relevant industrial and simulated examples.

Taming the Lion: Improving Plant Operations through Inferential Modeling

Conference: Southwest Process Technology Conference

Year: 2019

Proceeding: 11th AIChE Southwest Process Technology Conference

Group: Southwest Process Technology Conference

Session: IIoT/Big Data

Time: Tuesday, October 1, 2019 - 10:00am-10:30am

Authors:

Perez, K. X., Dow
Lien, R., Dow
Ashcraft, B., Dow
Gandhi, A., Dow

Inferential property models, often called soft-sensors, estimate process properties that cannot be measured directly or continuously using easily measured variables. Developed using various data analytic techniques, inferential property models enable improved control by providing estimated product properties in real-time. They are widely used in various chemical manufacturing processes at Dow. This presentation will provide some insights gained from the collaboration of soft-sensor development across functional teams in Dow and share implementation case studies and best practices on the use of inferential property models to improve plant operations.

From Artificial Intelligence to Augmented Intelligence

Conference: Southwest Process Technology Conference

Year: 2019

Proceeding: 11th AIChE Southwest Process Technology Conference

Group: Southwest Process Technology Conference

Session: IIoT/Big Data

Time: Tuesday, October 1, 2019 - 11:00am-11:30am

Authors:

Brigman, J., Ingenero Inc.

The use of artificial intelligence to process massive amounts of data in real time to analyze operations is considered a must-have for companies. Boards are pressuring CEOs to have a comprehensive digitalization strategy to take their companies into the future. Technology solutions are being adopted for their allure and hopes that additional novel uses will be found. These â€œsolutions-looking-for-a-problemâ€ are ultimately failing on the economic benefit test. Why are these projects getting to avoid the standard return on investment criteria that all other projects face? Fear of falling behind is not justification for dropping business sense. The ultimate goal of running a business is and should be to generate a profitable return.

Good ROI on big data / artificial intelligence is coming from those initiatives that have a problem to solve. In the process industry increased profits come from increased throughput, yield, and equipment availability without diminishing safety or the capital efficiency to achieve this higher production. This comes down to correctly answering the questions related these core areas and making the changes and investments to achieve maximum profitability. It is the use of artificial intelligence to augment the intelligence of the companyâ€™s personnel to make the right moves. Certainly automating the right moves is an ultimate goal. Automating is 20th centuryâ€¦that can be done. Providing the insight to make the right moves is the promise of artificial intelligence.

This presentation looks at specific use cases where artificial intelligence and machine learning should be utilized to augment the intelligence of operations engineers and operators to enable profitability enhancement through increased throughput, yield and equipment availability.

Is Artificial Intelligence (AI) Creating Value for Your Business? Improving Chemical Processes Using Iiot, Big Data and Digital Twins

Conference: Southwest Process Technology Conference

Year: 2019

Proceeding: 11th AIChE Southwest Process Technology Conference

Group: Southwest Process Technology Conference

Session: IIoT/Big Data

Time: Tuesday, October 1, 2019 - 11:30am-12:00pm

Authors:

Washburn, J., DeepIQ

You may already use smart sensor technology to enhance your chemical processes. Do you have huge quantities of data collected and stored in multiple formats? How much of this data can you use today?

In this session, we will explore how IIoT data can be easily stored and organized using a modern big data platform. We will cover how to simply prepare your data for analysis by exploiting the parallel computing power of big data clusters.

Finally, we will show how to create a digital representation of your physical assets such as compressors. Using a â€œDigital Twinâ€, we will show how you can run different models to meet your business needs such as improve production, optimize costs, or to predict failures.

Engineering Ethics

Conference: Southwest Process Technology Conference

Year: 2019

Proceeding: 11th AIChE Southwest Process Technology Conference

Group: Southwest Process Technology Conference

Session: Ethics Session

Time: Tuesday, October 1, 2019 - 12:30pm-12:55pm

Authors:

Harkness, M., AIChE South Texas Section

The Cambridge Dictionary defines Ethics as â€œa system of accepted beliefs that control behavior, especially such a system based on moralsâ€, and morals are defined as â€œâ€¦.standards of good behavior, honesty, and fair dealingâ€¦.â€. Engineering Ethics is the application of moral standards to the practice of Engineering, and defines an Engineerâ€™s responsibilities to and behavior toward clients, the engineering profession, and society.

Pressurized LNG - a stranded gas solution and a greener form of LNG

Conference: Southwest Process Technology Conference

Year: 2019

Proceeding: 11th AIChE Southwest Process Technology Conference

Group: Southwest Process Technology Conference

Session: Chemical Process Technology

Time: Tuesday, October 1, 2019 - 1:30pm-2:00pm

Authors:

White, C. N., ezNG Solutions LLC

Optimally balancing temperature with moderate pressure can significantly reduce the complexity and cost of producing a liquid phase of natural gas suitable for storage and transport. Currently over 650 BBTU/day are flared in the Permian Basin alone. Adopting this optimal practice to reducing the energy and waste involved in liquefaction brings a greener solution for disposal of associated gas when pipelines are not accessible. The facilities required for conditioning and chilling associated gas for production of conventional LNG requires unattractive levels of investment. Going with Pressurized LNG (PLNG) changes the game. A pioneering cost comparison between PLNG and LNG introduces ezNG Solutionsâ€™ technology as a means to profitably avoid the tragic waste and losses that burden the shale industry.

Pact™, a New Flexible and Robust Technology for the Conversion of Light Alkanes to Aromatics, Octane

Conference: Southwest Process Technology Conference

Year: 2019

Proceeding: 11th AIChE Southwest Process Technology Conference

Group: Southwest Process Technology Conference

Session: Chemical Process Technology

Time: Tuesday, October 1, 2019 - 2:00pm-2:30pm

Authors:

Baldridge, A., Phillips 66
Poulsen, J. M., Haldor Topsoe A/S

PACTâ„¢, a New Flexible and Robust Technology for the Conversion of Light Alkanes to Aromatics, Octane

Paraffin Activation and Conversion Technology (PACTâ„¢), jointly developed by Phillips66 and Haldor Topsoe is a new, flexible, and robust technology for the conversion of C5+/natural gasoline to higher valued products. The recent increase in the supply of natural gas liquids (NGLs) due to the U.S. shale revolution motivates an opportunity to upgrade surplus C5+/natural gasoline streams to products with larger market and higher value. Existing commercial technologies for NGL conversion are optimized for lighter paraffins, but more cost-effective upgrading of C5+/natural gasoline streams can be realized with novel advancements in catalysis coupled with a unique process design.

Current U.S. EIA estimates show a 60% increase in C5+/natural gasoline supply within the US from 2018 to 2022 with supply continuing to outpace demand. Additionally, global supplies of C5+/natural gasoline/light naphtha are expected to rise at similar rates providing a global overall surplus of these streams. While this supply continues to grow, traditional outlets for C5+/natural gasoline, including gasoline blending, diluent for heavy crudes, and naphtha steam cracking, are likely unable to absorb the growing surplus of C5+/natural gasoline.

The newly developed PACTâ„¢ Process converts C5+/natural gasoline/very light naphtha and other light alkane streams into more valuable products including high octane blendstocks, aromatics, olefins, and hydrogen through the utilization of well-proven and robust operation units to minimize technical risk. Pilot studies of the paraffin activation step of the PACT Process have demonstrated process flexibility for a variety of streams. This presentation will include pilot-scale results highlighting the yield and conversion structure along with key process and economic indicators for this new technology.

Utilizing Gamma Scans to Monitor Fouling Accumulation in Packed Towers

Conference: Southwest Process Technology Conference

Year: 2019

Proceeding: 11th AIChE Southwest Process Technology Conference

Group: Southwest Process Technology Conference

Session: Chemical Process Technology

Time: Tuesday, October 1, 2019 - 2:30pm-3:00pm

Authors:

Mixon, W., Tracerco

In some processes, a common problem is the accumulation of fouling material in the packed beds of separation towers. Fouling can cause conditions such as liquid and/or vapor maldistribution and excessive liquid holdup in packed columns, resulting in a loss of efficiency, an increase in operating pressure drop and premature flooding. Fouling may result from foreign material entering a column or from polymerization or decomposition occurring inside the column. The fouling may start under normal conditions or begin to build due to abnormal operating conditions.

During a turnaround, a visual inspection can be performed on the upper and lower layers of packing, but unless the packing is removed, it is impossible to see fouling material within the bed. In some cases, towers with fouled beds are restarted without cleaning or replacing the packing, because no fouling was visually seen on the upper or lower layers of packing during the inspection. This can ultimately lead to costly unplanned outages.

Gamma scanning is a non-intrusive and cost-effective way to monitor a packed tower for the accumulation of fouling material. Many operators implement a routine maintenance program where the column is periodically scanned to monitor the progress of the condition over time. This type of monitoring program typically begins with a baseline scan of the column.

This paper will provide case studies that illustrate the effectiveness of utilizing gamma scans to monitor fouling accumulation in packed towers.

Kbr's Propane Dehydrogenation Technology (K-PRO TM) Meets Market-Driven on-Purpose Propylene Demand

Conference: Southwest Process Technology Conference

Year: 2019

Proceeding: 11th AIChE Southwest Process Technology Conference

Group: Southwest Process Technology Conference

Session: Chemical Process Technology

Time: Tuesday, October 1, 2019 - 3:30pm-4:00pm

Authors:

Caton, J., KBR

Propylene is mostly produced worldwide as a by-product from either steam crackers or FCC units. However, with the large number of new ethane-based steam crackers coming on line in the next few years, the ongoing transition of many existing steam crackers from naphtha feedstock to ethane and propane feedstock, and with very few new FCC units being built, it is anticipated that these traditional sources of propylene supply will fall short of demand by 45 MMTPA in 2027. 80% of this propylene shortfall is expected to be made up by on-purpose propane dehydrogenation. KBR's Propane Dehydrogenation Technology (K-PRO ™), based on the commercially proven KBR Catalytic Olefins Technology (K-COT ™) in combination with a novel high selectivity dehydrogenation catalyst, is well positioned to fill this gap. The technology can be implemented as a stand-alone propylene production unit independent of a steam cracker or an FCC unit.

During this presentation, we will present the details of K-PRO ™Technology and discuss its benefits including high propylene selectivity and conversion, low-cost and environmentally friendly catalyst, minimal catalyst losses and catalyst make-up requirements, how its commercially proven reactor design leads to simple and reliable operation with high on-stream factors, and lower CAPEX and OPEX as compared to other technologies.

Olefin Plant Wastewater Treatment Issues – Unique Challenges and their Solutions

Conference: Southwest Process Technology Conference

Year: 2019

Proceeding: 11th AIChE Southwest Process Technology Conference

Group: Southwest Process Technology Conference

Session: Chemical Process Technology

Time: Tuesday, October 1, 2019 - 4:00pm-4:30pm

Authors:

Basu, S., Headworks International

Olefin Plant Wastewater Treatment Issues â€“ Unique Challenges and their Solutions

Presentation by: **Paul T Sun, PhD, PE and Somnath Basu, PhD, PE Headworks International**

ABSTRACT

The ethylene/propylene production industry has been profoundly changed due to the frac gas production. A large number of world class gas steam crackers are being designed or under construction in the Gulf Coast Region and beyond. Although the wastewater production of these units is not as complex compared with integrated refinery/naphtha cracker, it has unique characteristics which will demand careful evaluations. These issues are:

1. Spent caustic treatment
2. Quench water pretreatment
3. Cooling water handling
4. Biological wastewater treatment due to inorganic nutrient deficiency
5. Others

The presentation will consist of:

1. A general discussion of the ethane cracking process and wastewater generation;
2. Characteristics of effluent streams generated from gas cracking operations;
3. Challenges handling, treatment and management of gas cracking wastewater;
4. Wastewater treatment processes â€" physicochemical and biological
5. Case histories on the challenges and their mitigation at three crackers globally will be presented.

Optimizing Fractionation Condenser System CAPEX and OPEX—Case Studies & Performance

Conference: Southwest Process Technology Conference

Year: 2019

Proceeding: 11th AIChE Southwest Process Technology Conference

Group: Southwest Process Technology Conference

Session: Distillation & Separation

Time: Tuesday, October 1, 2019 - 1:30pm-2:00pm

Authors:

Parvin, E., Parv Consulting

Whether designing a new fractionation train or debottlenecking an existing one, there are opportunities in condenser systems that are often overlooked. This presentation will focus on two key concepts.

1). Fractionation systems often experience operational issues or bottlenecks caused by the condenser system, either by design or through creep capacity over time. Two case studies from refining and midstream industries will be shared where design of new condensers for fractionation systems considered overhead line pressure drop causing superheat entering the condenser. This superheat can impact sizing and/or performance of the condenser as well as the entire distillation system. Potential root causes for missing these design details in todayâ€™s industry will be discussed. One case study will demonstrate simultaneous reduction of both CAPEX and OPEX costs. These same considerations can be used to evaluate and debottleneck existing systems.

2) A case study will be presented from the design of a now operating facility for a feed/condenser heat integration scheme. Common arrangements of such an exchanger from both refining and midstream experiences will be covered. The best overall layout of an actual depropanizer system with key detailed design considerations to ensure operation success will be shared, which realized both CAPEX and OPEX savings for the project. Retrofitting existing system with this heat integration scheme may debottleneck condenser limited fractionation systems for higher capacity at relatively lower costs than traditional methods.

Topic Selection:

Debottlenecking Condenser Limited Distillation Systems with Novel Use of Reflux

Abstract Text:

Distillation columns when capacity is creeped over time, and in many cases upon startup of a newly designed facility, often experience bottlenecks on the overhead condensers. Methods to avoid bottlenecks of new designs will be presented briefly. For existing systems experiencing bottlenecks, we will present common root causes of problems of condenser limited systems, with a focus of overhead line pressure drop to the condenser and how it impacts the heat release curve performance of the condenser. Then a review of conventional methods of debottlenecking. Lastly a new novel and counter-intuitive method (patent pending) for using reflux in a new way to debottleneck the condenser limited systems in most distillation systems. Benefits of using this method may include but not be limited to: Increased capacity of the entire distillation system; lower operating pressure of the tower year-round; improved performance of the condenser; lower operating cost of the entire distillation system at the same fixed rates; and any combination of the above. Quantitative analysis of the heat release curve and performance of a typical tower system will be presented.

A copy of the patent submittal is available upon request to our office.

Presenting Author

Eric Parvin

Parv Consulting
Owner / Consulting Engineer

Details Matter when Designing Hot Vapor Bypass Systems

Conference: Southwest Process Technology Conference

Year: 2019

Proceeding: 11th AIChE Southwest Process Technology Conference

Group: Southwest Process Technology Conference

Session: Distillation & Separation

Time: Tuesday, October 1, 2019 - 2:00pm-2:30pm

Authors:

Hanson, D., Valero Energy Corp.

Pressure control is the most important parameter that is used in distillation optimization. Engineers realize that there are three important parameters in distillation optimization: temperature, pressure, and composition. When two of these are controlled, the other parameter is fixed. When pressure and temperature are controlled and/or optimized, the composition is known. Effective column pressure control is critical to controlling product quality targets.

Hot Vapor Bypass systems have been used extensively in the industry. There have been several good papers recently regarding trouble areas of the industryâ€™s design methodology. This paper will illustrate a severe instability that occurred within Valero.

This case study will illustrate that even when a design configuration has been installed many times; it can be the source of problems, unexpected consequences, and even possibly result in unit shutdowns. In this case study severe disturbances resulted. This design has been used many times in industry. Design practitioners should consider this experience and abandon this type of design that are prone to operating instabilities. Good designs are those that avoids failure even when it is operating outside of its intended operation range.

A Non-traditional Approach that Improves Random Packing Performance

Conference: Southwest Process Technology Conference

Year: 2019

Proceeding: 11th AIChE Southwest Process Technology Conference

Group: Southwest Process Technology Conference

Session: Distillation & Separation

Time: Tuesday, October 1, 2019 - 2:30pm-3:00pm

Authors:

Garcia, A., AMACS

â€œA Non-traditional Approach that Improves Random Packing Performance.â€

Antonio Garcia

ABSTRACT

The chemical process industry benefits greatly from any improvements on mass transfer equipment performance due to its impact on energy consumption in distillation processes containing conventional equipment -trays and/or packings-. In existing grass-roots plants, process intensification (higher throughputs, higher purities, or yields) is typically achieved by upgrading from conventional to high-performance equipment and optimizing operating conditions. In high pressure distillation columns designed with trays, these columns are upgraded to contain either high-capacity trays or high-performance random packings. The latter typically achieved using fourth-generation random packings with improved geometries. These types of packings exhibit better hydraulics leading to higher capacities.

The patented non-traditional approach has been available for many years. This approach consists in blending different sizes of a third-generation random packing widely used in the industry. The resulting performance of the blended packing has a similar capacity to the larger random packing size with an improved efficiency.

Recent experimental work conducted in independent research centers confirm the improved hydraulic and mass transfer performance of this non-traditional approach.

A well-known design methodology for random packed columns is successfully applied when modeling these random packing blends.

Rapid Evaluation of FCC Main Fractionator using HYSYS Column Analyzer

Conference: Southwest Process Technology Conference

Year: 2019

Proceeding: 11th AIChE Southwest Process Technology Conference

Group: Southwest Process Technology Conference

Session: Distillation & Separation

Time: Tuesday, October 1, 2019 - 3:30pm-4:00pm

Authors:

Robertson, E., Burns and McDonnell

Rapid Evaluation of FCC Main Fractionator using HYSYS Column Analyzer

Elliott Robertson

Burns & McDonnell

This paper summarizes rapid evaluation of an existing FCC Main Fractionator using HYSYS Column Analyzer. A HYSYS V9 model was developed for the Main Frac and was tuned against test run data at maximum stable operating rates. The Main Frac was known to flood at a certain feed and LCO sidedraw rates. Gamma scans indicated flooding in the LCO draw section, and HYSYS Column Analyzer was used to confirm these results. Replacing trays in the existing Main Frac with packing for pressure drop reduction did not alleviate tower flooding, as shown by Column Analyzer. Reduction of steam and slurry rates to minimum levels did not alleviate flooding. A replacement tower diameter was determined with HYSYS Column Analyzer in interactive sizing mode. A new tower would allow more separation stages allowing better LCO quality increasing product value which results in more favorable payback. HYSYS Column Analyzer was used to demonstrate operational flexibility with the new tower with respect to feed changes and seasonal effects.

Unexpected Problems with Kettle Reboiler Circuit

Conference: Southwest Process Technology Conference

Year: 2019

Proceeding: 11th AIChE Southwest Process Technology Conference

Group: Southwest Process Technology Conference

Session: Distillation & Separation

Time: Tuesday, October 1, 2019 - 4:00pm-4:30pm

Authors:

Cai, T., Fractionation Research, Inc.
Cavett, M., Fractionation Research, Inc.

Unexpected Problems with Kettle Reboiler Circuit

Mark Cavett, Tony Cai*

Fractionation Research, Inc.

ABSTRACT

Reboilers are responsible for many operational problems experienced in distillation and absorption columns in the chemical processing industry.Employed in a commercial size distillation research facility, Fractionation Research, Inc (FRI) kettle reboiler circuits operate under a wide range of process conditions and various physical properties. This makes their operation more challenging. In this paper, the unexpected operational challenges and problems encountered with the kettle reboiler circuits will be described and discussed. The troubleshooting process, solutions to the operational problems, and the lesson learned will be presented.

Key Words: Distillation, reboiler, hydraulics, troubleshooting

Expect the Unexpected: Hydrogen Sulfide Exposure during Troubleshooting

Conference: Southwest Process Technology Conference

Year: 2019

Proceeding: 11th AIChE Southwest Process Technology Conference

Group: Southwest Process Technology Conference

Session: Process Safety

Time: Tuesday, October 1, 2019 - 1:30pm-2:00pm

Authors:

Pearson, K., Upstream Engineering Centre

During routine troubleshooting work involving multi-stage vacuum jets, an unexpected release of Hydrogen Sulfide (H2S) at the indoor job site led to exposure of two people, one of whom later died. The worker who subsequently died was attending to his coworker, whom he believed was suffering from a heart attack. Several additional individuals were exposed when they rushed to the room without protective equipment; fortunately, all made a full recovery.

This paper discusses the circumstances that led to this tragic incident as well as the management system and cultural issues that contributed to the event and to the subsequent emergency response. Key learnings included:

- Troubleshooting and maintenance activities require robust pre-job planning and hazard recognition processes.
- Hazardous by-products can be generated in a facility, even from non-hazardous chemicals, and can result in exposures unless well understood. This facility was not subject to the OSHA 1910.119 PSM regulation, and the raw materials and products were non-hazardous.
- A strong Incident Command structure can reduce risk to personnel during emergency response.

Note that this incident did not occur at a facility owned by BP Americas at the time of the incident. However, the speaker was employed by the company that owned this facility at the time of the incident and she was the lead investigator

Process Safety and Our Next Generation Workforce

Conference: Southwest Process Technology Conference

Year: 2019

Proceeding: 11th AIChE Southwest Process Technology Conference

Group: Southwest Process Technology Conference

Session: Process Safety

Time: Tuesday, October 1, 2019 - 2:00pm-2:30pm

Authors:

Brister, D., Shea Capability & Compliance Solutions, LLC

The skills gap between the current, aging workforce and the next generation of workers is well documented. But what may be less widely known is how generation gaps can affect workplace safety and skills gaps for our next generation workforce and the frontline leaders, particularly when new workers have fewer capable and seasoned personnel working alongside themâ €"people who could share their practical safety and technical knowledge. This lack of one-on-one support can, and does, result in more workplace incidents, damaged equipment, wasted resources and time. The problem is compounded when leaders prioritize results without reinforcing safe work practices and operational processes. As a result, younger, less experienced workers feel they must get the job done without really understanding how to best mitigate risk, or attempt a task without fully understanding what they are undertaking.

Process Safety has just started to really get embedded in the workplace here in the U.S. since its rollout as law in 1992. Now, those with all the process safety knowledge have either left your company, getting ready to leave or will be leaving in the next decade.

- What has your company done to capture your tribal process safety knowledge and -practices?
- Who will be there to guide the hearts and minds of a much younger and inexperienced workforce?
- What are you doing now to manage this risk right now?

Weâ€™ll show what the experts in the field of human learning and develop say, especially for those who must have the required skills working in high hazard environments where knowing and understanding process safety is critical to safe and reliable operations, regardless how much technology you have in place to minimize human error.

Keywords: safety culture, competency, leadership, next generation workforce

Integrated PHA & LOPA Review Guideline to improve the SIL Classification Process

Conference: Southwest Process Technology Conference

Year: 2019

Proceeding: 11th AIChE Southwest Process Technology Conference

Group: Southwest Process Technology Conference

Session: Process Safety

Time: Tuesday, October 1, 2019 - 2:30pm-3:00pm

Authors:

Clark, T., Bechtel Oil, Gas & Chemicals, Inc.
Girada, R., Bechtel Oil, Gas & Chemicals, Inc.
Balakrishnan, S., Bechtel Oil, Gas & Chemicals, Inc.

The integration of layer of protection analysis (LOPA) with process hazard analysis (PHA) conducted using the hazard and operability (HAZOP) methodology has become the best approach to Functional Safety Management. Until a few years ago, Safety Integrity Level (SIL) assessment study in the form of integrated LOPA was not always conducted in quick succession to the PHA study, or with the same team, during the Front-End Engineering Design (FEED) phase of a typical LNG or petrochemical facility. This has often led to incorrect allocation of required risk reduction by a Safety Instrumented Function (SIF).

Experience has confirmed that having the same team participating in integrated PHA and LOPA reviews has improved the effectiveness of both studies by:

- Eliminating analysis of the same hazard-consequence scenarios twice,
- Bringing consistency with risk assessed previously in HAZOP reviews and later SIL level classification (or assignment),
- Having fewer optimized preventive or mitigative actions compared to performing only a HAZOP, and
- Faster support of control systems engineering with comprehensive preparation of the Safety Requirement Specifications (SRS).

This presentation details the concept/guidelines of integrating LOPA with PHA using the

HAZOP methodology. This approach has been internally adopted in Bechtel Oil, Gas & Chemicals, Inc. across several projects, achieving efficient and consistent results.

Hazard Identification and Risk Analysis: What Can Go Wrong?

Conference: Southwest Process Technology Conference

Year: 2019

Proceeding: 11th AIChE Southwest Process Technology Conference

Group: Southwest Process Technology Conference

Session: Process Safety

Time: Tuesday, October 1, 2019 - 3:30pm-4:00pm

Authors:

Hart, R. J., Exponent Inc.
Morris, J., Exponent Inc.

Hazard Identification and Risk Analysis is an encompassing term for activities involved in identifying hazards and evaluating their risk at a facility. Risk to employees, the public, and the environment all need to be controlled within a stakeholderâ€™s acceptable risk tolerance. A study to evaluate the risk encompasses the following three areas 1) Hazardâ€"What can go wrong? 2) Consequenceâ€"How bad can it be? and 3) Likelihoodâ€"How often will it happen? To manage risk, a quantitative or qualitative risk analysis identifies the potential risks which are then evaluated and compared against risk criteria to determine tolerability. This presentation will cover common methods for performing Hazard Identification, Consequence Analysis, Likelihood Determination, and the resulting Risk Assignment. Pros and cons of the various techniques will be discussed, with examples of case studies (including LNG liquefaction plants and transportation projects) where the various techniques may be suitable.

Properly Size Two-Phase PRV - Easier Than API 520 HDI Method

Conference: Southwest Process Technology Conference

Year: 2019

Proceeding: 11th AIChE Southwest Process Technology Conference

Group: Southwest Process Technology Conference

Session: Process Safety

Time: Tuesday, October 1, 2019 - 4:00pm-4:30pm

Authors:

Chen, G., Enerflex

API 520 Part I (9th edition) selected three (3) methods to size two-phase PRV. One of them is the HDI (Homogeneous Direct Integration) method described in section C.2.1. HDI method involves generating multiple data points over an isentropic range of pressure from the inlet to the discharge. These data are used to evaluate the mass flux integral by direct numerical integration.

The HDI method requires intensive effort depending on how close engineers desire the result, comparing to the actual integration. Therefore, an easier HD (Homogeneous Direct) method without integration, yet producing identical mass flux, is needed and proposed.

The new easier HD (Homogeneous Direct) method compares the velocity of fluid and the velocity of sound along the isentropic expansion path. The choke point is determined when the velocity of fluid equals the velocity of sound. The mass flux is then simply calculated by multiplying mass density and the velocity of fluid at the choke point. A detailed example will be given in the presentation.

The HD method is compared against the established data in â€œBenchmarking of two-phase flow through safety relief valves and pipesâ€ by Shawn Adair and Harold Fisher. The mass flow rate through a 4P6 relief valve by HD method is well within the reasonable range, among other established methods. The new easier HD method is also validated against API examples for gas, liquid, two-phase and supercritical relief.

Evidence of Climate Change: An Overview of the Science

Conference: Southwest Process Technology Conference

Year: 2019

Proceeding: 11th AIChE Southwest Process Technology Conference

Group: Southwest Process Technology Conference

Session: Climate Solutions

Time: Wednesday, October 2, 2019 - 8:15am-8:45am

Authors:

Thomas, S., NA

Global temperature is a function of the balance between incoming and outgoing energy. An increase in greenhouse gas emissions has altered that balance, leading to global warming. Since 1880, global temperature has risen 0.8 Â°C[1]. This presentation will describe the major drivers of global climate, document the many lines of evidence for the increase in global temperature, and provide information about the implications of atmospheric and oceanic warming.

Warming will impact the Southern Great Plains (Texas, Oklahoma, and Kansas), a major hub for energy production, export, and freight. According to the 4th National Climate Assessment (2018),[2] annual average temperatures for this region are projected to increase by 3.6Â°â€“5.1Â°F by the mid-21st century. This region will likely face costly impacts due to more intense and frequent extreme weather events like heat, drought, flooding and storms. These extreme events are expected to make severe demands on aging infrastructure and stress on our regionâ€™s water resources. Furthermore, rising sea levels will place coastal infrastructure at risk.

[1] https://earthobservatory.nasa.gov/world-of-change/DecadalTemp

[2] https://nca2018.globalchange.gov/chapter/23/

Strategic Approaches in Developing and Implementing Mitigating Solutions

Conference: Southwest Process Technology Conference

Year: 2019

Proceeding: 11th AIChE Southwest Process Technology Conference

Group: Southwest Process Technology Conference

Session: Climate Solutions

Time: Wednesday, October 2, 2019 - 8:45am-9:15am

Authors:

Quazi, H., AngelNet Energy, Water and Environmental Systems Inc

The presentation will highlight the strategic approaches in developing and implementing solutions for oil refining, gas processing, chemical and petrochemical plants. It will provide certain examples of appropriate programs for mitigating greenhouse-gases (GHGs) in these plants. Also, comments will be shared relative to addressing â€œenergy, water and foods nexusâ€ related challenges during the first-half of the 21st Century.

Risky Business? The Very Tangible Supply Chain Threats of Climate Change

Conference: Southwest Process Technology Conference

Year: 2019

Proceeding: 11th AIChE Southwest Process Technology Conference

Group: Southwest Process Technology Conference

Session: Climate Solutions

Time: Wednesday, October 2, 2019 - 9:45am-10:15am

Authors:

Berg, M., Simfero Consultants

Industry consolidation, customer demands, and competition from abroad present complex challenges that continue to ratchet up pressures on the chemical industry. Climate change threatens this precarious balance even further. Yet the vast majority of climate change discussions are framed in terms of the distant future. While conditions will indeed lead to increasing impacts over time, we donâ€™t have to wait decades. In many places, these impacts are already being felt, with critical implications for chemical engineering supply chains. This presentation will take a close look at current temperature and water resources trends in selected locations, from source to delivery. Special attention will be given to impacts at facilities in Texas. To ensure durability of both individual organizations and the industry as a whole, it will only become more crucial to respond to climate trends at both global and on-site levels.

Climate Resilience: Translating Facility Preparedness to Solutions for Today and Tomorrow

Conference: Southwest Process Technology Conference

Year: 2019

Proceeding: 11th AIChE Southwest Process Technology Conference

Group: Southwest Process Technology Conference

Session: Climate Solutions

Time: Wednesday, October 2, 2019 - 10:15am-10:45am

Authors:

Luman, R. L., AECOM

Science-based solutions and engineering principals are (thankfully!) addressing the global, long-term effects of human-induced climate change. Recognizing that coastal regions support valuable eco-systems and play a critical role in energy security and local and global economies, requires industry, the government, and communities to act ***now;***collaborating to identify and implement practical solutions, creating resilient coastal communities that protect our economy and our environment now and into the future.

Historical, facility-specific vulnerability assessments and disaster plans have not considered the interconnectivity of industry, community, and the environment. With increasing instances of extreme weather events, facility-specific planning requires reassessment to prepare for the myriad challenges brought by such extreme events disrupting industry and their connected communities. This presentation focuses on the Texas Gulf Coast and the importance of climate resiliency planning and preparedness***now***, auditing facilities through a new lens to make informed, targeted investments for the site and beyond, acting collaboratively to implement solutions that provide multiple values to minimize business interruption and protect our critical coastal regions from physical and environmental impacts today and tomorrow.

Through the first wave: What's next for US capital investments

Conference: Southwest Process Technology Conference

Year: 2019

Proceeding: 11th AIChE Southwest Process Technology Conference

Group: Southwest Process Technology Conference

Session: Petroleum Refining Technology

Time: Wednesday, October 2, 2019 - 8:15am-8:45am

Authors:

Nichols, L., Hydrocarbon Processing / Gas Processing

The increase in shale oil and gas production has led to significant capital investments in the US midstream and downstream industries. This buildout includes billions of dollars in new refining and petrochemical capacity, pipeline infrastructure and LNG and oil export terminals. This presentation will explore capital-intensive investments in the midstream and downstream sectors of the US, the significance of this buildout, and what lies ahead for the countryâ€™s capital investments.

Enterprise Wide Fired Heater Fleet Monitoring Tool

Conference: Southwest Process Technology Conference

Year: 2019

Proceeding: 11th AIChE Southwest Process Technology Conference

Group: Southwest Process Technology Conference

Session: Petroleum Refining Technology

Time: Wednesday, October 2, 2019 - 8:45am-9:15am

Authors:

Trivedi, K. K., ExxonMobil
Kubis, E., Exxonmobil
Solomon, C., Exxonmobil
Nur Hidayah, M. N., Exxonmobil
Chunangad, K., Exxonmobil
Nehete, S., Exxonmobil
Gunter, J., Exxonmobil

Enterprise Wide Fired Heater Fleet Monitoring Tool

Kirtan Trivedi[1], Ed Kubis[2], Cameron Solomon[3], Md Nasir Nur Hidayah[4], Kris Chunangad[5], Suhas Nehete[4], John Gunter[6]

1. ExxonMobil Global Projects Company
2. ExxonMobil Chemical Company
3. ExxonMobil Fuels and Lubricant Company, Joliet Refinery
4. ExxonMobil Research and Engineering Company.
5. ExxonMobil Business Support Center, Malaysia
6. ExxonMobil Fuels & Lubricants Company

In response to challenges related to global warming, climate change, and greenhouse gas emissions, several countries are considering to adopt regulatory measures requiring industry to further improve energy efficiency. ExxonMobil has a long tradition of effectively managing energy efficiency, both for internal operational efficiency, as well as in support of external drivers such as the API Compendium challenge within the industry. Over several decades, substantial work has been done from an energy management perspective. This has enabled ExxonMobil to become the most energy efficient international refining company in the world and the most energy efficient refining company operating in the U.S. per Solomon Associates benchmarking. The company has achieved a 10 percent improvement in energy efficiency across its global refining operations following an effort launched in 2000 [1]. However, with emerging regulations across various countries, itâ€™s becoming increasingly important to develop smart ways to identify further energy efficiency improvement opportunities in existing facilities.

Fired Heaters is one of the largest sources of energy input for a refinery. Ensuring efficient operation of fired heaters will help improve energy efficiency of existing facilities. The in-house tool developed by the authors, leverages data analytics principles and data visualization programs to identify actionable opportunities. This minimizes engineering resources required for identification of efficiency improvement projects. Performance issues are identified in a timely manner, improving long term global fleet efficiency trends. Opportunities are identified based on historical data and technically achievable targets. Global priorities and business plans are set using this tool to close the gaps for restoring and improving performance of the fleet. This single platform helps to develop internal benchmarks with improved learning around fleet metrics.

This paper discusses how the tool is developed, its functionality, and examples on how the tool is used to monitor the fleet, and identify opportunities for efficiency improvement.

Reference:

- https://news.exxonmobil.com/press-release/exxonmobil-announces-greenhouse-gas-reduction-measures; accessed 7/17/2019

Digitalization: Assuring Your Plant Achieves Its Full Potential1

Conference: Southwest Process Technology Conference

Year: 2019

Proceeding: 11th AIChE Southwest Process Technology Conference

Group: Southwest Process Technology Conference

Session: Petroleum Refining Technology

Time: Wednesday, October 2, 2019 - 9:15am-9:45am

Authors:

Williams, C., KBC

Presenting the results from the digitalization proof of concept that commenced late 2017 at a leading US refinery.

The proof of concept has used cloud technology to bring together data, leading rigorous simulation, latest generation analytics and global subject matter experts. With the aim of assuring that the unit maintains its full potential; multi-million dollarsâ€™ worth of opportunities have been identified to date.

Plants understand the value of rigorous simulation, but most simulators are not an up to date representation of the plant or do not connect to historian for their â€œAs-Isâ€ data. Therefore, plant staff spend valuable hours in reconciling simulations versus actual versus planning and manually updating reports, calculations, KPIs, etc. from simulation re-runs. Consequently, investigation of plant excursions can be slow and result in delaying corrective actions.

Latest tools in digitalization has allowed the integration of data and simulation systems to create a plant digital twin that automatically updates as plant values changes. This allows for an accurate representation of the asset over its full range of operation and to capture the full history and future of the asset. In addition, SMEs supporting the local process engineers proactively troubleshoot the unitâ€™s bottlenecks and continuously offer knowledge transfer. The result is a centralized single version of the truth, used by everyone, outputs delivered directly to the business, strong governance systems and a unit that runs at maximum economical potential.

The presentation will detail the study methodology and describe the collaborative digital solutions that connects the refinery to the industryâ€™s global talent. Such as:

- Plant data and unit economics through a Digital Mirror
- Continuously rating unit performances with real time reconciled data and a cloud-based unit model
- â€˜Badâ€™ data avoidance with educated analytic conditioning alongside determination of instrumentation or process problems
- Planning modelâ€™s validity checks against the changing operating window
- Current profitability and gap to best economic potential dashboards
- Expert recommendations and stewardship actions to close economic gap

The paper concludes with a critical examination of the learning from the study implementation and how the traditional work processes of the leading refinery adapted to the digitalization in order to achieve the increase in profitability.

High Performance Column Internals for Fouling Applications

Conference: Southwest Process Technology Conference

Year: 2019

Proceeding: 11th AIChE Southwest Process Technology Conference

Group: Southwest Process Technology Conference

Session: Petroleum Refining Technology

Time: Wednesday, October 2, 2019 - 9:45am-10:15am

Authors:

Pilling, M. W., Sulzer Chemtech USA, Inc.

Fouling applications are found throughout the refining and chemical processing industries. This causes additional cost to plant operations due to subsequent limitations in capacity and/or efficiency and nearly always reduces the allowable unit run length prior to shutdown for cleaning. From an equipment design standpoint, distillation and absorption columns typically use larger orifices and hydraulic forces to keep fouling materials moving. Unfortunately, such modifications often limit the capacity and efficiency of the associated internals. Sulzer has developed several tray and packing options that are proven to resist fouling while maintaining high performance. This presentation will discuss fouling applications and detail the operating characteristics of these high performance fouling resistant devices and design solutions.

October 2, 2019

Effective Interviewing

Building the Foundation for Personal and Professional Success

Greg Yeo
Chief Engineer, ExxonMobil Chemical Company

11th AIChE Southwest Process Technology Conference, Sugarland

---

What is Effective Interviewing?

- **Advance preparation:** Preparing as if it's an important class
  - Understand the position, understand yourself
  - Prepare to tell your story in a relevant and compelling way
- **Presenting yourself at the Interview**
  - Being authentic, relevant and motivated
- **Self-evaluation**
  - How did I do?
- **Improve**
  - Building skills, accomplishments, credibility and a career

The Interviewer

- What is the interviewer looking for? *The "right" balance of:*
  - Technical Ability
  - Leadership
  - Collaboration
  - Communication
  - Grit
- What does the interviewer have to go on?
  - Academic record
  - Resume / CV
  - *Prior reference (maybe)?*
  - *Internship (maybe?)*
  - *Short Campus interview (maybe?)*
  - *Phone interview (maybe?)*
  - *Recommendations?*

"Does this person make my team better? Is this an engineer that I want on my next project?"

ExxonMobil

Preparation

<u>Action</u>

- Learn about the company and role in advance.
  - What do they do?
  - What are the roles?
  - Local or global?
  - Technology or design or manufacturing or research or sales or…
    - o What is the necessary balance of Technical Ability, Leadership, Collaboration, Communication

<u>Insight</u>

Very few students do this

Technical Leadership Collaboration Communication Grit	vs.	Technical Leadership Collaboration Communication Grit

ExxonMobil

Preparation

Action

- Prepare an interesting resume that is a platform to tell your story.
  - Technical capability
  - Leadership
  - Collaboration
  - Grit

Insight

Very few students do this

- Each element of your resume is an opportunity to tell your story
- Your interview is your demonstration of Communication skill
  - Presence
  - Belonging
  - Warmth

ExxonMobil

Preparation

Action

- Become GOOD at telling your story.
  - Relevant
  - Interesting
  - Credible
  - Respectful
- Become good at having a professional conversation. *Practice!*

Insight

Very few students do this

- Be genuine. Do not be scripted.
- Listen to good communicators. *What do they do that makes them good?*
  - Presence
  - Belonging
  - Warmth
- Practice. Record yourself, then listen to it.

ExxonMobil

Preparation

Action

- Stretch yourself. Seek out opportunities to grow your skills.
  - Technical
  - Leadership
  - Collaboration
  - Communication
  - Grit

Insight

- Start now. Lean in to every opportunity to grow and improve.
- ***Develop these skills:***
  - While in school
  - Work
  - Volunteer / non-profit
  - Sports
  - Arts
- Find & grow mentors / role models in these critical areas. *The fundamentals are much more important than tricks & tips & techniques*

ExxonMobil

If You Have A Pre-interview:

- This is part of the interview process – it matters
  - Be yourself! This person is trying to get to know you.
  - Are you the kind of person they would want to work with?
- Ask questions
  - Learn. Show genuine interest. Make this part of your prep.
  - This person got a job – learn from them!
- Be energetic, interested, professional

Before I interview a student,
I always talk to the person who talked to them in advance.

ExxonMobil

At the Interview

- Be enthusiastic, engaging, and **authentic**.
  - Make eye contact. Show interest. And mean it.
- Create a two way conversation
  - You are interested in this interviewer and company
  - Ask questions. Take the lead when it's offered.
  - ***Be able to tell your interesting story that relates to the position / company.***
- Align the conversation with your story as it relates to the job
  - *Technical depth*
  - *Leadership*
  - *Working with / connecting with people*
  - *Ability to stick with tough challenges*

ExxonMobil

Be Thoughtful. Be Professional. Be Influential.

- Credibility = Demonstrated Competence + Integrity + Relationships
- Be careful & thoughtful with social media
- Build bridges, build a career
- Focus on getting better with continued learning:
  - Challenges & issues in need of solutions
  - Technical advances & innovations
  - Business / Industries / Companies
  - Cultures
  - Leadership

ExxonMobil

Modularization & Intensification of Carbon Capture Technologies

Conference: Southwest Process Technology Conference

Year: 2019

Proceeding: 11th AIChE Southwest Process Technology Conference

Group: Southwest Process Technology Conference

Session: Process Intensification

Time: Wednesday, October 2, 2019 - 1:30pm-2:00pm

Authors:

De Leon, R., University of Houston
Datta, A., University of Houston
Krishnamoorti, R., University of Houston

Carbon capture technologies that capture CO_2 from point sources such as power plants, refineries, and chemical plants or distributed and typically low concentration sources like the atmosphere are being advanced as part of a comprehensive carbon management system. Fundamentally, there are three pathways to capture point source generated CO_2. They depend on when and how the CO_2 is captured in the combustion process: pre-combustion, post-combustion, and oxy-fuel combustion carbon capture. These processes have been scaled up to minimize the energy required for releasing the CO_2 and for operations including pipeline compression of CO_2. Such point source capture technologies have demonstrated improvements in energy efficiency through the integration of processes and more recently by application of intensification methods. On the other hand, direct air capture (DAC) methods involve low concentration streams, are intrinsically smaller scale, and are distributed. DAC proves economical when adopting passive technologies to capture CO_2. In this presentation, we will provide techno-economic analysis of existing DAC technologies and life cycle analysis to understand the efficacy of these methods to reduce the global carbon footprint. We will also discuss the technical opportunities to modularize and intensify such distributed capture technologies to address energy consumption, high capex costs, and integration of renewable energy sources to provide an alternate pathway for rapid penetration of carbon capture technologies.

How modern digital design approaches can help realize the potential of Process Intensification

Conference: Southwest Process Technology Conference

Year: 2019

Proceeding: 11th AIChE Southwest Process Technology Conference

Group: Southwest Process Technology Conference

Session: Process Intensification

Time: Wednesday, October 2, 2019 - 2:00pm-2:30pm

Authors:

Leyland, S., Process Systems Enterprise Inc

Despite the numerous potential benefits, the process industry has been slow in the adoption of Process Intensification since its inception. A key challenge is that intensified processes are by definition novel and unproven, as opposed to less-efficient processes that have been well-understood for many years and therefore carry less risk. A traditional approach to process developments dictates that new processes require extensive construction of prototypes and pilots. However, even exhaustive pilot testing still leaves open questions of operability and reliability, and a lack of systematic quantification of the effects of poor performance or failure, as well as the usual general technology risks associated with implementing new processes. There is also a perceived lack of design tools and data to develop intensified processes, and lack of generalized workflows for dealing with the complexity of intensified, integrated modular systems.

All of this means that significant advantages can be realized from applying emerging digital design approaches that allow rapid and systematic exploration of the process decision space and rigorous quantification and management of technology risk. Digital design employs a model-based approach coupled closely with targeted experimentation. Experimentation is used to support the construction of a high-fidelity predictive model (â€˜digital twinâ€™); once a model of sufficient accuracy is established, the digital twin, rather than the experimental data, is used to optimize the process design and operation.

This presentation describes, with brief illustrations, the established digital design techniques, technologies, and workflows that can be applied across the intensified process development lifecycle to accelerate development and manage risk systematically.

Operability challenges in Process Intensification – the role of dynamic modelling & advanced control

Conference: Southwest Process Technology Conference

Year: 2019

Proceeding: 11th AIChE Southwest Process Technology Conference

Group: Southwest Process Technology Conference

Session: Process Intensification

Time: Wednesday, October 2, 2019 - 2:30pm-3:00pm

Authors:

Bindlish, R., Dow Chemical
Pistikopoulos, S., Texas A&M University

In this presentation, we highlight the importance of considering operability criteria in process intensification. In particular, we discuss and analyze the role of key factors such as degrees of freedom, constraints, dynamics and modular/equipment size in the context of process intensification and in comparison with conventional systems; and how dynamic modelling and advanced control can help us towards the development and deployment of computational tools for the design of operable intensified process systems.

AUTHOR INDEX